Contents

What is spying 4
Targets of espionage 5
Methods and terminology 6
Technology and techniques 8
Industrial espionage 11
Agents in espionage 12
Law 15
History of espionage laws 16
Use against non-spies 17
Espionage laws in the UK 18
Military conflicts 20
Spy fiction 22
12 Self-Reliance Tools You Should Own 24
The Spy Secret to Being Invisible 28
How to Expose a Tail 35
The Worst Places to Hide Your Valuables 39
Invasion of the Privacy-Snatchers 42
What is a fusion center 42
To Catch a Terrorist 43
The Plot Thickens 44
A Spy's Eye View 45

Keeping an Eye on Things 46
In the Eyes of the Law 48

What is spying

Espionage or spying is the act of obtaining secret or confidential information without the permission of the holder of the information. Spies help agencies uncover secret information. Any individual or spy ring (a cooperating group of spies), in the service of a government, company or independent operation, can commit espionage. The practice is clandestine, as it is by definition unwelcome. In some circumstances it may be a legal tool of law enforcement and in others it may be illegaland punishable by law. Espionage is a method of intelligence gathering which includes information gathering from non-disclosed sources.

Espionage is often part of an institutional effort by a government or commercial concern. However, the term tends to be associated with state spying on potential or actual enemies for military purposes. Spying involving corporations is known as industrial espionage.

One of the most effective ways to gather data and information about a targeted organization is by infiltrating its ranks. This is the job of the spy (espionage agent). Spies can then return information such as the size and strength of enemy forces. They can also find dissidents within the organization and influence them to provide further information or to defect.[2] In times of crisis, spies steal technology and sabotage the enemy in various ways. Counterintelligence is the practice of thwarting enemy espionage and intelligence-gathering. Almost all nations have strict laws concerning espionage and the penalty for being caught is often severe. However, the benefits gained through espionage are often so great that most governments and many large corporations make use of it.

Information collection techniques used in the conduct of clandestine human intelligence include operational techniques, asset recruiting, and tradecraft.

Today, espionage agencies target the illegal drug trade and terrorists as well as state actors. Since 2008, the United States has charged at least 57 defendants for attempting to spy for China.

Intelligence services value certain intelligence collection techniques over others. The former Soviet Union, for example, preferred human sources over research in open sources, while the United States has tended to emphasize technological methods such as SIGINT and IMINT. In the Soviet Union, both political (KGB) and military intelligence (GRU[4]) officers were judged by the number of agents they recruited.

Targets of espionage

Espionage agents are usually trained experts in a targeted field so they can differentiate mundane information from targets of value to their own organizational development. Correct identification of the target at its execution is the sole purpose of the espionage operation

Broad areas of espionage targeting expertise include:

Natural resources: strategic production identification and assessment (food, energy, materials). Agents are usually found among bureaucrats who administer these resources in their own countries

Popular sentiment towards domestic and foreign policies (popular, middle class, elites). Agents often recruited from field journalistic crews, exchange postgraduate students and sociology researchers

Strategic economic strengths (production, research, manufacture, infrastructure). Agents recruited from science and technology academia, commercial enterprises, and more rarely from among military technologists

Military capability intelligence (offensive, defensive, manoeuvre, naval, air, space). Agents are trained by military espionage education facilities and posted to an area of operation with covert identities to minimize prosecution

Counterintelligence operations targeting opponents' intelligence services themselves, such as breaching the confidentiality of communications, and recruiting defectors or moles

Methods and terminology

Although the news media may speak of "spy satellites" and the like, espionage is not a synonym for all intelligence-gathering disciplines. It is a specific form of human source intelligence (HUMINT). Codebreaking (cryptanalysis or COMINT), aircraft or satellite photography, (IMINT) and research in open publications (OSINT) are all intelligence gathering disciplines, but none of them is considered espionage. Many HUMINT activities, such as prisoner interrogation, reports from military reconnaissance patrols and from diplomats, etc., are not considered espionage. Espionage is the disclosure of sensitive information (classified) to people who are not cleared for that information or access to that sensitive information.

Unlike other forms of intelligence collection disciplines, espionage usually involves accessing the place where the desired information is stored or accessing the people who know the information and will divulge it through some kind of subterfuge. There are exceptions to physical meetings, such as the Oslo Report, or the insistence of Robert Hanssen in never meeting the people who bought his information.

The US defines espionage towards itself as "The act of obtaining, delivering, transmitting, communicating, or receiving information about the national defence with an intent, or reason to believe, that the information may be used to the injury of the United States or to the advantage of any foreign nation". Black's Law Dictionary (1990) defines espionage as: "... gathering, transmitting, or losing ... information related to the national defense". Espionage is a violation of United States law, 18 U.S.C. §§ 792–798 and Article 106a of the Uniform Code of Military Justice".[5] The United States, like most nations, conducts espionage against other nations, under the control of the National

Clandestine Service. Britain's espionage activities are controlled by the Secret Intelligence Service.

Technology and techniques

• Agent handling

• Concealment device

• Covert agent

• Covert listening device

• Cut-out

• Cyber spying

• Dead drop

• False flag operations

• Honeypot

• Impersonation

• Impostor

• Interrogation

• Non-official cover

• Numbers messaging

• Official cover

• One-way voice link

- Sabotage

- Safe house

- Side channel attack

- Steganography

- Surveillance

- Surveillance aircraft

Organization

An intelligence officer's clothing, accessories, and behaviour must be as unremarkable as possible — their lives (and others') may depend on it.

A spy is a person employed to seek out top secret information from a source. Within the United States Intelligence Community, "asset" is more common usage. A case officer or Special Agent, who may have diplomatic status (i.e., official cover or non-official cover), supports and directs the human collector. Cutouts are couriers who do not know the agent or case officer but transfer messages. A safe house is a refuge for spies. Spies often seek to obtain secret information from another source.

In larger networks, the organization can be complex with many methods to avoid detection, including clandestine cell systems. Often the players have never met. Case officers are stationed in foreign countries to recruit and to supervise intelligence agents, who in turn spy on targets in their countries where they are assigned. A spy need not be a citizen of the target country—hence does not automatically commit treason when operating within it. While the more common practice is to recruit a person already trusted with access to sensitive information, sometimes a person with a well-prepared synthetic identity (cover background), called a legend in tradecraft, may attempt to infiltrate a target organization.

These agents can be moles (who are recruited before they get access to secrets), defectors (who are recruited after they get access to secrets and leave their country) or defectors in place (who get access but do not leave).

A legend is also employed for an individual who is not an illegal agent, but is an ordinary citizen who is "relocated", for example, a "protected witness". Nevertheless, such a non-agent very likely will also have a case officer who will act as a controller. As in most, if not all synthetic identity schemes, for whatever purpose (illegal or legal), the assistance of a controller is required.

Spies may also be used to spread disinformation in the organization in which they are planted, such as giving false reports about their country's military movements, or about a competing company's ability to bring a product to market. Spies may be given other roles that also require infiltration, such as sabotage.

Many governments spy on their allies as well as their enemies, although they typically maintain a policy of not commenting on this. Governments also employ private companies to collect information on their behalf such as SCG International Risk, International Intelligence Limited and others.

Many organizations, both national and non-national, conduct espionage operations. It should not be assumed that espionage is always directed at the most secret operations of a target country. National and terrorist organizations and other groups are also targeted.[citation needed] This is because governments want to retrieve information that they can use to be proactive in protecting their nation from potential terrorist attacks.

Communications both are necessary to espionage and clandestine operations, and also a great vulnerability when the adversary has sophisticated SIGINT detection and interception capability. Agents must also transfer money securely

Industrial espionage

Reportedly Canada is losing $12 billion[7] and German companies are estimated to be losing about €50 billion ($87 billion) and 30,000 jobs[8] to industrial espionage every year.

Agents in espionage

In espionage jargon, an "agent" is the person who does the spying; a citizen of one country who is recruited by a second country to spy on or work against his own country or a third country. In popular usage, this term is often erroneously applied to a member of an intelligence service who recruits and handles agents; in espionage, such a person is referred to as an intelligence officer, intelligence operative or case officer. There are several types of agent in use today:

• Double agent: "engages in clandestine activity for two intelligence or security services (or more in joint operations), who provides information about one or about each to the other, and who wittingly withholds significant information from one on the instructions of the other or is unwittingly manipulated by one so that significant facts are withheld from the adversary. Peddlers, fabricators and others who work for themselves rather than a service are not double agents because they are not agents. The fact that double agents have an agent relationship with both sides distinguishes them from penetrations, who normally are placed with the target service in a staff or officer capacity."[9]

• Redoubled agent: forced to mislead the foreign intelligence service after being caught as a double agent.

• Unwitting double agent: offers or is forced to recruit as a double or redoubled agent and in the process is recruited by either a third-party intelligence service or his own government without the knowledge of the intended target intelligence service or the agent. This can be useful in capturing important information from an agent that is attempting to seek allegiance with another country. The double agent usually has knowledge of both intelligence services and can identify operational techniques of both, thus making third-party recruitment difficult or impossible. The knowledge of operational techniques can also affect the relationship between the operations officer (or case officer) and the agent if the case is transferred by an operational targeting officer to a new operations officer, leaving the new officer vulnerable to attack. This type of transfer may occur when an officer has completed his term of service or when his cover is blown.

• Triple agent: works for three intelligence services.

• Intelligence agent: provides access to sensitive information through the use of special privileges. If used in corporate intelligence gathering, this may include gathering information of a corporate business venture or stock portfolio. In economic intelligence, "Economic Analysts may use their specialized skills to analyze and interpret economic trends and developments, assess and track foreign financial activities, and develop new econometric and modelling methodologies."[10] This may also include information of trade or tariff.

• Access agent: provides access to other potential agents by providing profiling information that can help lead to recruitment into an intelligence service.

• Agent of influence: provides political influence in an area of interest, possibly including publications needed to further an intelligence service agenda. The use of the media to print a story to mislead a foreign service into action, exposing their operations while under surveillance.

• Agent provocateur: instigates trouble or provides information to gather as many people as possible into one location for an arrest.

• Facilities agent: provides access to buildings, such as garages or offices used for staging operations, resupply, etc.

• Principal agent: functions as a handler for an established network of agents, usually considered "blue chip."

• Confusion agent: provides misleading information to an enemy intelligence service or attempts to discredit the operations of the target in an operation.

• Sleeper agent: recruited to wake up and perform a specific set of tasks or functions while living undercover in an area of interest. This type of agent is not the same as a deep cover operative, who continually contacts a case officer to file intelligence reports. A sleeper agent is not in contact with anyone until activated.

- Illegal agent: lives in another country under false credentials and does not report to a local station. A nonofficial cover operative can be dubbed an "illegal"[11] when working in another country without diplomatic protection.

Law

Espionage is a crime under the legal code of many nations. In the United States, it is covered by the Espionage Act of 1917. The risks of espionage vary. A spy breaking the host country's laws may be deported, imprisoned, or even executed. A spy breaking their own country's laws can be imprisoned for espionage or/and treason (which in the US and some other jurisdictions can only occur if they take up arms or aids the enemy against their own country during wartime), or even executed, as the Rosenbergs were. For example, when Aldrich Ames handed a stack of dossiers of U.S. Central Intelligence Agency (CIA) agents in the Eastern Bloc to his KGB-officer "handler", the KGB "rolled up" several networks, and at least ten people were secretly shot. When Ames was arrested by the U.S. Federal Bureau of Investigation (FBI), he faced life in prison; his contact, who had diplomatic immunity, was declared persona non grata and taken to the airport. Ames' wife was threatened with life imprisonment if her husband did not cooperate; he did, and she was given a five-year sentence. Hugh Francis Redmond, a CIA officer in China, spent nineteen years in a Chinese prison for espionage—and died there—as he was operating without diplomatic cover and immunity.[12]

In United States law, treason,[13] espionage,[14] and spying[15] are separate crimes. Treason and espionage have graduated punishment levels.

The United States in World War I passed the Espionage Act of 1917. Over the years, many spies, such as the Soble spy ring, Robert Lee Johnson, the Rosenberg ring, Aldrich Hazen Ames,[16] Robert Philip Hanssen,[17] Jonathan Pollard, John Anthony Walker, James Hall III, and others have been prosecuted under this law.

History of espionage laws

From ancient times, the penalty for espionage in many countries was execution. This was true right up until the era of World War II; for example, Josef Jakobs was a Nazi spy who parachuted into Great Britain in 1941 and was executed for espionage.

In modern times, many people convicted of espionage have been given penal sentences rather than execution. For example, Aldrich Hazen Ames is an American CIA analyst, turned KGB mole, who was convicted of espionage in 1994; he is serving a life sentence without the possibility of parole in the high-security Allenwood U.S. Penitentiary.[18] Ames was formerly a 31-year CIA counterintelligence officer and analyst who committed espionage against his country by spying for the Soviet Unionand Russia.[19] So far as it is known, Ames compromised the second-largest number of CIA agents, second only to Robert Hanssen, who is also serving a prison sentence.

Use against non-spies

Espionage laws are also used to prosecute non-spies. In the United States, the Espionage Act of 1917 was used against socialist politician Eugene V. Debs (at that time the Act had much stricter guidelines and amongst other things banned speech against military recruiting). The law was later used to suppress publication of periodicals, for example of Father Coughlin in World War II. In the early 21st century, the act was used to prosecute whistleblowers such as Thomas Andrews Drake, John Kiriakou, and Edward Snowden, as well as officials who communicated with journalists for innocuous reasons, such as Stephen Jin-Woo Kim.

As of 2012, India and Pakistan were holding several hundred prisoners of each other's country for minor violations like trespass or visa overstay, often with accusations of espionage attached. Some of these include cases where Pakistan and India both deny citizenship to these people, leaving them stateless. The BBC reported in 2012 on one such case, that of Mohammed Idrees, who was held under Indian police control for approximately 13 years for overstaying his 15-day visa by 2–3 days after seeing his ill parents in 1999. Much of the 13 years were spent in prison waiting for a hearing, and more time was spent homeless or living with generous families. The Indian People's Union for Civil Liberties and Human Rights Law

Network both decried his treatment. The BBC attributed some of the problems to tensions caused by the Kashmir conflict.

Espionage laws in the UK

Espionage is illegal in the UK under the Official Secrets Acts of 1911 and 1920. The UK law under this legislation considers espionage as "concerning those who intend to help an enemy and deliberately harm the security of the nation". According to MI5, a person commits the offence of 'spying' if they, "for any purpose prejudicial to the safety or interests of the State": approaches, enters or inspects a prohibited area; makes documents such as plans that are intended, calculated, or could directly or indirectly be of use to an enemy; or "obtains, collects, records, or publishes, or communicates to any other person any secret official code word, or password, or any sketch, plan, model, article, or note, or other document which is calculated to be or might be or is intended to be directly or indirectly useful to an enemy". The illegality of espionage also includes any action which may be considered 'preparatory to' spying, or encouraging or aiding another to spy.

An individual convicted of espionage can be imprisoned for up to 14 years in the UK, although multiple sentences can be issued.

Government intelligence laws and its distinction from espionage

Government intelligence is very much distinct from espionage, and is not illegal in the UK, providing that the organisations of individuals are registered, often with the ICO, and are acting within the restrictions of the Regulation of Investigatory Powers Act (RIPA). 'Intelligence' is considered legally as "information of all sorts gathered by a government or organisation to guide its decisions. It includes information that may be both public and private, obtained from much different public or secret sources. It could consist entirely of information from either publicly available or secret sources, or be a combination of the two."

However, espionage and intelligence can be linked. According to the MI5 website, "foreign intelligence officers acting in the UK under diplomatic cover may enjoy immunity from prosecution. Such persons can only be tried for spying (or, indeed, any criminal offence) if diplomatic immunity is waived beforehand. Those officers operating without diplomatic cover have no such immunity from prosecution".

There are also laws surrounding government and organisational intelligence and surveillance. Generally, the body involved should be issued with some form of warrant or permission from the government and should be enacting their procedures in the interest of protecting national security or the safety of public citizens. Those carrying out intelligence missions should act within not only RIPA but also the Data Protection Act and Human Rights Act. However, there are spy equipment laws and legal requirements around intelligence methods that vary for each form of intelligence enacted.

Military conflicts.

In military conflicts, espionage is considered permissible as many nations recognize the inevitability of opposing sides seeking intelligence each about the dispositions of the other. To make the mission easier and successful, soldiers or agents wear disguises to conceal their true identity from the enemy while penetrating enemy lines for intelligence gathering. However, if they are caught behind enemy lines in disguises, they are not entitled to prisoner-of-war status and subject to prosecution and punishment—including execution.

The Hague Convention of 1907 addresses the status of wartime spies, specifically within "Laws and Customs of War on Land" (Hague IV); October 18, 1907: CHAPTER II Spies".[25] Article 29 states that a person is considered a spy who, acts clandestinely or on false pretences, infiltrates enemy lines with the intention of acquiring intelligence about the enemy and communicate it to the belligerent during times of war. Soldiers who penetrate enemy lines in proper uniforms for the purpose of acquiring intelligence are not considered spies but are lawful combatants entitled to be treated as prisoners of war upon capture by the enemy. Article 30 states that a spy captured behind enemy lines may only be punished following a trial. However, Article 31 provides that if a spy successfully rejoined his own military and is then captured by the enemy as a lawful combatant, he cannot be punished for his previous acts of espionage and must be treated as a prisoner of war. Note that this provision does not apply to citizens who committed treason against their own country or co-belligerents of that country and may be captured and prosecuted at any place or any time regardless whether he rejoined the military to which he belongs or not or during or after the war.

The ones that are excluded from being treated as spies while behind enemy lines are escaping prisoners of war and downed airmen as international law distinguishes between a disguised spy and a disguised escaper. It is permissible for these groups to wear enemy uniforms or civilian clothes in order to facilitate their escape back to friendly lines so long as they do not attack enemy forces, collect military intelligence, or engage in similar military operations while so disguised. Soldiers who are wearing enemy uniforms or civilian clothes simply for the sake of warmth along with other purposes rather than engaging in espionage or similar military operations while so attired are also excluded from being treated as unlawful combatants.

Saboteurs are treated as spies as they too wear disguises behind enemy lines for the purpose of waging destruction on an enemy's vital targets in addition to intelligence gathering. For example, during World War II, eight German agents entered the U.S. in June 1942 as part of Operation Pastorius, a sabotage mission against U.S. economic targets. Two weeks later, all were arrested in civilian clothes by the FBI thanks to two German agents betraying the mission to the U.S. Under the Hague Convention of 1907, these Germans were classified as spies and tried by a military tribunal in Washington D.C. On August 3, 1942, all eight were found guilty and sentenced to death. Five days later, six were executed by electric chair at the District of Columbia jail. Two who had given evidence against the others had their sentences reduced by President Franklin D. Roosevelt to prison terms. In 1948, they were released by President Harry S. Truman and deported to the American Zone of occupied Germany.

The U.S. codification of enemy spies is Article 106 of the Uniform Code of Military Justice. This provides a mandatory death sentence if a person captured in the act is proven to be "lurking as a spy or acting as a spy in or about any place, vessel, or aircraft, within the control or jurisdiction of any of the armed forces, or in or about any shipyard, any manufacturing or industrial plant, or any other place or institution engaged in work in aid of the prosecution of the war by the United States, or elsewhere".[

Spy fiction

Spies have long been favourite topics for novelists and filmmakers.[34] An early example of espionage literature is Kim by the English novelist Rudyard Kipling, with a description of the training of an intelligence agent in the Great Game between the UK and Russia in 19th century Central Asia. Even earlier work was James Fenimore

Cooper's classic novel, The Spy, written in 1821, about an American spy in New York during the Revolutionary War.

During the many 20th-century spy scandals, much information became publicly known about national spy agencies and dozens of real-life secret agents. These sensational stories piqued public interest in a profession largely off-limits to human interest news reporting, a natural consequence of the secrecy inherent in their work. To fill in the blanks, the popular conception of the secret agent has been formed largely by 20th and 21st-century fiction and film. Attractive and sociable real-life agents such as Valerie Plame find little employment in serious fiction, however. The fictional secret agent is more often a loner, sometimes amoral—an existential hero operating outside the everyday constraints of society. Loner spy personalities may have been a stereotype of convenience for authors who already knew how to write loner private investigator characters that sold well from the 1920s to the present.

Johnny Fedora achieved popularity as a fictional agent of early Cold War espionage, but James Bond is the most commercially successful of the many spy characters created by intelligence insiders during that struggle. His less fantastic rivals include Le Carre's George Smiley and Harry Palmer as played by Michael Caine.

Jumping on the spy bandwagon, other writers also started writing about spy fiction featuring female spies as protagonists, such as The Baroness, which has more graphic action and sex, as compared to other novels featuring male protagonists.

It also made its way into the videogame world, hence the famous creation of Hideo Kojima, the Metal Gear Solid Series.

Espionage has also made its way into comedy depictions. The 1960s TV series Get Smart portrays an inept spy, while the 1985 movie Spies Like Us depicts a pair of none-too-bright men sent to the Soviet Union to investigate a missile.

12 Self-Reliance Tools You Should Own

If a crisis situation chases us out of our homes, even temporarily, we'll be better off than most people if we've prepared a bug-out bag with nonperishable food and other essential items, as well as a water purifier.

But if we're forced to stay away from our homes for an extended period of time, we will also benefit greatly by having certain tools in our vehicles. In fact, those tools could make the difference between life and death.

Let's take a look at 12 of the most important ones:

Survival Knife — This self-reliance tool is practical in many situations and could keep you alive if things turn ugly. It should be sturdy and lightweight with a single fixed blade. Plan on spending a couple hundred dollars on this item.

Multitool — A must-have in your survival supplies but also in your emergency car survival kit. Most multitools will contain pliers, folding knife, small saw, metal file, hole punch, screwdriver, wood saw, wire cutter and even a scissors. They range in price from $15 to $130.

Hatchet — Make sure you pick something that's lightweight and versatile. Expert campers and wilderness survival experts recommend hatchets over axes. There are many uses for this dependable tool. Test the sharpness before adding it to your emergency supplies and make sure it includes a sheath.

Fire Starter Kit — In an emergency situation, you'll want the ability to quickly produce fire not just for heat but also to aid in emergency meal preparation. You can purchase an emergency fire kit or create your own. Either way, make sure it's waterproof, easy to carry and includes a light for making fire in the dark.

Emergency Radio — There are two things you'll do first in a catastrophe. Seek safety and then seek information. Emergency radios come in all shapes and sizes. Most people will only need a radio to hear NOAA alerts and other warnings. Before you spend a lot of money on two-way or shortwave radios, consider whether or not you'll need them.

Stainless Steel Canteen — Get one with a screw-on lid, designed for hiking. It will carry water and can also be used to boil and/or purify water for drinking from questionable sources.

Signaling Equipment — Signaling devices such as flares and glow sticks should be included in an emergency kit. A loud whistle is also recommended, as you never know when you might end up trapped during a disaster. A high-pitched whistle is likely to catch someone's attention and break through background noise.

Compass — With so many options from which to choose, make sure you have a small and very durable compass and aren't planning on relying solely on GPS. The most important aspect is to possess the knowledge of how to use whichever compass you purchase. Practice is key.

Emergency Blanket — A durable, compact and lightweight emergency survival blanket will reflect heat back to your body, protect you from the elements and possibly serve as shelter. Include several, as they are a necessity in all types of emergency situations.

Headlamp — Not having light during an emergency can be frustrating, demoralizing and, most importantly, dangerous. This is why you should have multiple light sources in your emergency supplies. Hopefully, you already have emergency candles, high-powered LED flashlights and a lantern in your survival supplies, but include a headlamp as well.

Duct Tape — This is an inexpensive addition that will come in handy in many instances. This is the time to channel your inner MacGyver and realize just how handy duct tape can be during an emergency. You can seal doors and windows and even safely remove glass shards from broken windows.

Knowledge — Lastly, the most important self-reliance tool to have is your personal survival knowledge. The more you know about surviving during emergency conditions and how to properly use your survival supplies, the more likely you are to stay alive.

Of course, it's important to remember that in a crisis, your No. 1 need is food, and that should be the first item you take care of for your family.

In fact, experts say that everyone needs to have nonperishable, good-for-25-years survival food on hand in case of an emergency.

Well, right now, for a limited time, we're giving away free 72-hour Food4Patriots survival food kits to loyal subscribers as long as they beat the program deadline (and while supplies last).

Every 72-hour kit that's being given away contains 16 total servings of such delicious meals as Blue Ribbon Creamy Chicken Rice, the always-loved Granny's Home Style Potato Soup and stick-to-your-ribs breakfast favorite Maple Grove Oatmeal.

This kit sells to the general public for $27 plus postage and has been rated 4.5 out of 5 stars by customers.

But readers who act quickly can get them just for the shipping-and-handling fee.

There is still time to take advantage of this offer, but be aware that supplies are limited and the program may end at any time.

The Spy Secret to Being Invisible

It was nearly impossible to narrow down this week's batch of must-read articles. In the end, I decided not to limit myself to five pieces. Instead, I elected to include a sixth bonus article for your reading pleasure.

1. These Verbal Weapons Can Keep You Safer Than Any Gun

Imagine being able to stop a mugger in their tracks with the mere power of your voice. Or fooling a thief into stealing junk and leaving you your valuables with a cleverly delivered joke. Or, better yet, thwarting a carjacking with seven simple words.

Not only is it possible, it's easy. Click on the link above to find out how you can become a mighty wordsmith, capable of protecting yourself with a single phrase.

2. New Report Says Tech Companies Spy on Students in School

It's an entirely different world out there than it was when I was a kid. There are many parenting challenges that simply didn't exist 30, 20, even 10 years ago — especially when it comes to technology.

Technology is being integrated into classrooms more and more but without proper security protections — which is putting your children at risk. According to this article, the Electronic Frontier Foundation (EFF), a leading nonprofit organization defending civil liberties in the digital world, "investigated 152 tech services currently used in classrooms and found they 'were lacking in encryption, data retention and data sharing policies.'"

This means that your child's personal information — including their email address, photos, interests, even birth date — could easily be monitored, sold or hacked. So talk to your kids about what apps they're downloading and what information they're providing. Teach them to read and understand privacy policies. It's never too early to start putting good cybersecurity habits into practice.

3. How to Be the Gray Man When SHTF

The concept of the "gray man" is a person (man or woman) who strives to blend into their surroundings so as to be completely forgettable — as if they were never even there. This tactic is particularly useful in emergency scenarios.

This piece from Graywolf Survival does a great job of breaking down the theory of the gray man and explaining how to become invisible yourself if the SHTF. Take a look and be sure to click on the link to the supplemental article "Five Gray Man Secrets I Learned as a Surveillance Operative" for more specific advice on how to disappear in a crowd.

4. FBI Documents Detail How the Russians Try to Recruit Spies

Remember Chi Mak? He's the Power Paragon employee who was convicted of acting as an unregistered agent of a foreign government and exporting U.S. military secrets to China. He was arrested in 2005, but he had been living and working in the United States since the late 1970s.

What about Donald Heathfield and Tracey Foley? This couple was arrested in 2010 on suspicion of being part of a Russian espionage ring tasked with gathering information on nuclear weapons, American policy toward Iran, CIA leadership and congressional politics, among other topics. They immigrated to the U.S. via Canada and spent over 20 years building up their cover stories.

The point here is that there is a distinct method by which foreign powers attempt to target, recruit and manipulate sources to steal state secrets. And it might seem like it's straight out of a Hollywood thriller, but sometimes the most unbelievable stories are actually true.

5. How to Grow Vegetables in Containers — Growing Food in Small Spaces

This piece offers some more practical prepping advice for apartment dwellers. If you've got some extra space, a balcony or a fire escape, you should consider growing fresh vegetables and herbs in containers. It's a good, inexpensive way to augment your food stores in case of an emergency.

It's also a great way to encourage healthy eating. And the healthier you are, the better you'll be able to weather an unexpected crisis. Not only that, but fresh food is just plain delicious!

6. Four Warning Signs of a Tornado and How to Prepare

We are approaching the middle of "tornado season," and according to weather reports, this year it's off to a very active start. Tornadoes are one of the more formidable natural disasters, because they seem to come out of nowhere without warning.

But if you know what to watch out for, you can buy yourself precious extra minutes to get to safety. This chapter gives you four signs of a potential touchdown as well as four other ways to stay informed ahead of the storm. Plus, it clarifies the important difference between a tornado watch and a tornado warning.

If you live in a tornado-prone area, please take a moment to read this and review your preparations. The last thing you want is to be scrambling for cover when the wind picks up.

See Who's at the Door — Even When You're Not at Home

In these pages I've discussed many ways to fortify your home — from home security systems to canine sentries to devices that make it look like someone is home. Today, I'm going to focus on a key area of home protection that is often overlooked — even though it's right in front of your face.

That's right, we're going to talk about your front door.

Over 33% of burglars gain entry to a home through the front door, whether it's unlocked, picked open or kicked in. And one of the most common ways criminals case a house is by ringing the doorbell to see if anyone is home.

Ding Dong, Who's There?

Take the case of James Goldman, for example. A few months ago, James was away from home when he received a notification on his cellphone from his home security system alerting him that someone was ringing the doorbell at his home.

As another security measure, James had installed a doorbell that included a camera, so he was able to see the person standing at the door. James didn't recognize the man and wasn't expecting any visitors.

The suspicious man turned around and left, only to return a few minutes later wearing a pair of heavy-duty gloves. He climbed onto the roof, at which point James began yelling at the strange man through the speaker on the doorbell, saying that he was calling the police.

The would-be thief hopped down from the roof and took off. James was able to prevent his home from being burgled by having a doorbell camera as part of his home security system, which allowed him to scare off the criminal.

If you haven't done so, I recommend looking into a doorbell security camera. Hopefully, you've already installed security cameras around your home, but having an additional camera on your doorbell gives you the ability to communicate with whomever is at your door.

Here are three security doorbells I suggest looking into:

SkyBell HD — The SkyBell HD Wi-Fi Video Doorbell offers 1080p HD-quality video with a 5x zoom. It allows you to monitor your front door anytime — not just when someone rings the bell. With the SkyBell, you can see and hear visitors at your door with no monthly subscription fee. The only drawback is that the SkyBell must be wired into your existing doorbell — it doesn't use a battery. This device sells for around $160 on Amazon.

Ring Video Doorbell — The Ring Wi-Fi-Enabled Video Doorbell is one of the most popular smart doorbells on the market. It can be wired into your existing doorbell or operate on a battery. Using your cellphone or tablet, you can easily view and speak with visitors at your door — even when you aren't home. However, Ring does charge a

$3 monthly fee to save and share your videos. It sells for around $200 on Amazon.

August Doorbell Cam — August also offers smart locks as well as doorbell cameras. If you are looking for both, this is a doorbell camera to consider. You can integrate the locks and the camera so you can essentially buzz someone in by unlocking your door after vetting them via the camera. The only drawback is that the August camera doesn't have night vision, so the picture quality at night isn't the best. The doorbell camera by itself sells for $190 on Amazon.

I believe you should continually evaluate your home defense plan to find ways to improve and strengthen it, which is why I recommend one of these doorbell cameras.

But that's not all. Here are a few other things you can do make your front door a hard target:

Install Schlage or Medeco locks — these locks are much more secure than their commonplace Kwikset counterparts, which are easy to pick

I also recommend having a reliable deadbolt from either Schlage or Medeco

Get a solid-core door — one that's made out of sturdy, solid materials like wood, steel or iron

Resist the temptation to have a decorative glass front door — a criminal could smash through this in seconds

Use a door barricade — like the ones made by Nightlock — to make it impossible for someone to kick in your front door.

Statistics show that front doors are prime targets, so anything you can do to help deter or identify a thief is worth adding to your home defense plan.

How to Expose a Tail

In a recent Spy & Survival Briefingalert, I outlined three steps to securing your wireless network from any unauthorized use.

There are several good reasons to do this. First, insecure networks are inconvenient, because the more devices there are hooked up to a particular network, the slower the internet will run.

Second, and much more importantly, an unsecured Wi-Fi network puts you at a greater risk of being targeted by hackers trying to steal your personal information.

But there's another justification — pointed out by one of my readers — for following the precautions I laid out to beef up your wireless security.

Take it away, James.

Any browsing done by unauthorized users is tied to your IP address. You don't want a pedophile hijacking your signal, because it can be traced back to you. Explain that to the local prosecutor...

— James W.

You bring up a great point, James. You have no idea what illegal behavior people could be engaging in when they're piggybacking on your Wi-Fi signal. The last thing you want is law enforcement knocking on your door and questioning you about your browsing history.

How do you spot a tail or a tag-team tail, and how do you evade it without looking like you are?

— Warren W.

If only one person is following you, eventually, it will be pretty easy to figure it out. If you keep seeing the same person or the same vehicle while you are out and about in public, after a few stops, it will become clear that you are being followed.

However, when there are multiple people involved in a tail, it's much more difficult to spot. Professionals will often trade off being the "eye" — the one who has a visual of the person they are following. For example, if you stop to get gas, your pursuers will probably switch off so it doesn't look like someone is following you into the gas station.

With that being said, there are a few simple things you can do to see if someone is following you without being obvious. First, it's important to act as natural as possible. Don't constantly stop to look around or glance behind you.

If you are on foot, one way to look around is to be social with others. For instance, if you pass someone walking their dog, you could ask to pet it. This gives you a chance to stop and look around while you are saying hello to man's best friend. It's a more natural stop that won't make you look paranoid.

If you are being followed in your car, one trick to expose a tail is to go down a one-way street. This forces the people following you down the one-way street as well, which will be pretty noticeable.

Now, I've just given you a 30,000-foot overview. Spies spend weeks upon weeks training on this stuff. But if the person following you is just your average criminal, it shouldn't be too difficult to give them the slip.

So which do you think will come first: the financial collapse or the EMP attack? Or perhaps both will occur simultaneously?

— A.W.

Well, A.W., I'm no fortune-teller, but my guess is that we will see another financial collapse before we see a large-scale EMP attack.

The truth is many people believe that the U.S. economy is headed for a cliff and that we could see some major setbacks in the next few years. Any number of factors — both at home and abroad — could send our economy into a free fall. For example, the fact that our national debt is out of control — and still climbing — doesn't bode well for the future.

I'm seeing widespread use of silencers on the hunting channels lately (e.g., outdoor and sportsman). What are the steps necessary in owning them? Thanks.

— Keith T.

Gun suppressors are becoming increasingly common among hunters. Obviously, it's a huge advantage if you can take a shot without scaring off the game. However, currently, it's a lengthy process to buy one.

First, you need to find a local approval, (FFL) Class 3 dealer, so you can fill out the paperwork for the suppressor you want to purchase. You will need to provide passport photos and fingerprint cards, so be sure you have everything ready before you go.

Once you fill out the application, you have to send it to the ATF (Bureau of Alcohol, Tobacco, Firearms and Explosives) along with a $200 check. Then you have to sit back and wait for approval. Once your request is granted, you can go back to the FFL dealer and pick up your suppressor. Fair warning, the approval process can take a few months at least, so don't expect a quick turnaround.

I am interested in learning self-defense skills that are appropriate for a 71-year-old woman who is mobile and in good health except for some soreness and stiffness. What do you recommend?

— Ann M.

I suggest sticking with simple techniques. What I mean is try to find a local self-defense instructor that teaches basic self-defense moves. I don't recommend trying to learn a specific type of martial arts, because it can take years to master enough skills to be effective.

Also, if you're a subscriber to my Spy & Survival Briefing, take a look through past issues. I've covered several simple self-defense moves that will work for people of all ages, highlighting exactly where to strike an attacker for maximum damage with minimal effort.

The Worst Places to Hide Your Valuables

During a home burglary, the majority of criminals will immediately head for the master bedroom and then work their way to other areas of the home. Burglars know that most people don't hide their valuables, so they'll head straight for jewelry boxes, dresser drawers, nightstands and closets, where important items are usually kept.

In fact, just last month, burglars broke into a Pinecrest, Florida, home and stole $100,000 worth of jewelry from the master bedroom.

It all began when Chandresh and Bhairavi Lakhani left their home to enjoy a night out on the town. Upon returning to their residence, Bhairavi entered the master bedroom. She was shocked to see the room had been ransacked and the thieves had stolen the couple's watch collection, which was valued at $100,000.

Since we know where criminals are likely to look, it's important to consider hiding your valuables in other areas of your home. Most break-ins transpire in 10 minutes or less, so the more difficult you make it for thieves to find the goods, the better your chances of thwarting their efforts.

With that being said, I realize there are homes of different sizes and different layouts, so I decided to share with you the five worst places to hide your valuables, so you can make the necessary changes depending on your situation:

Underwear/sock drawer — People assume socks and underwear are of no value to a thief. Unfortunately, this age-old thinking means criminals head right for this drawer. While most people would never want to touch another person's socks or underwear, criminals have no qualms about rifling through your intimate apparel for valuables.

Under the mattress — If you're like me, this is where you hid your most prized possessions when you were a kid. I remember stashing my BB guns underneath my mattress. (By the way, I'm pretty sure my mom found them easily.) Never hide anything under the mattress because — I assure you — criminals will take the extra two seconds to flip the mattress and check.

Your children's bedroom — Many years ago, this was a great place to hide valuables because criminals weren't exactly looking for toys to steal. However, nowadays kids have tablets, smartphones, gaming consoles and other electronics. So if a criminal searches the master bedroom to no avail, they will most likely hit the next bedroom down the hall.

In the freezer — This is a hiding spot made popular in the movies (and by corrupt politicians). Well, life imitates art, and the truth is a lot of people do hide money in their freezer. I've known people who even wrap their money in foil and write "frozen meat" on the foil. But criminals watch movies too, and they will check among your frozen foods. Also, I'm sure you've heard about burglars who like to eat while burglarizing a home. As weird as it sounds, it's not uncommon, since many burglars are desperate.

In the toilet tank — At first, the toilet tank seems like a good idea. However, this is another hiding place that Hollywood has ruined by showing people stuffing money, jewelry or other items into plastic bags and hiding them in the tank.

Hopefully, this list has made you rethink where to hide your valuables. If you currently use any of the places I mentioned above, I encourage you to move those items right away.

Instead, hide your valuables in places that are very inconvenient to access. For example, in your attic in a box marked "summer clothes" underneath a stack of other boxes. If it's a pain for you to get to your valuables, then it's a lot less likely a criminal will be able to find them.

Of course, you could also buy a large safe and bolt it to the ground, which is a good idea too.

Invasion of the Privacy-Snatchers

According to the Department of Homeland Security, there are currently 78 government fusion centers throughout the United States. You'll find at least one in each state and in most major cities.

What is a fusion center

The government defines a fusion center as "a collaborative effort of two or more agencies that provide resources, expertise and information to the center with the goal of maximizing their ability to detect, prevent, investigate and respond to criminal and terrorist activity."

Put more simply, these centers are a way for local, state and federal agencies to share intelligence information for analysis. However, recent events might make you wonder whether these centers are collecting information on everyday Americans who are not criminals.

It's a complicated issue — and there are people with strong points of view on both sides. Whether you support these centers or think they are an invasion of privacy, the truth is fusion centers have helped to prevent terrorist attacks and even catch everyday criminals.

To Catch a Terrorist

In 2009, one of these centers helped shut down a jihadist group in Raleigh, North Carolina. In the early 1990s, Daniel Patrick Boyd, an American citizen, trained and fought with militant groups in Afghanistan. Back in the States, he began recruiting members to join a group of men to commit deadly terrorist attacks around the world.

Boyd enlisted his two sons, along with several other men, to join his group. On July 27, 2009, Boyd and seven men were arrested for conspiring to carry out attacks overseas, where they planned to kidnap and murder innocent people, in addition to engaging in violent jihad.

At this point, you're probably asking yourself how a fusion center factored into the unraveling of this nefarious plot.

According to investigators, someone tipped off the community outreach branch of the fusion center in North Carolina. The valuable information was shared among federal and state agencies and helped authorities build a case against the terror group.

This is just one example of how the cooperation fostered by fusion centers has foiled terrorists' plans — and that's not all. Fusion centers have also helped catch serial bank robbers, drug smugglers and arms dealers.

But the main focus of these centers, according to the government, is to effectively communicate between organizations such as the CIA, the FBI, the U.S. Department of Justice, the U.S. military and state and local law enforcement. Fusion centers collect the raw information from these different agencies and use it to analyze any perceived threat.

What does this mean for you?

Well, it's important to note that many of these centers monitor social media platforms like Facebook, Twitter and Instagram — particularly the accounts of people or groups who organize protests. Fusion centers across the country have been keeping tabs on environmental activists, Second Amendment groups, tea partiers and Occupy Wall Street protesters, among others.

As I mentioned, many people disagree with the idea of fusion centers because they believe these centers are invading the privacy of innocent Americans, while others question their effectiveness.

The Plot Thickens

According to a 2012 Senate report, the Department of Homeland Security had spent as much as $1.4 billion funding fusion centers since 2003, but the subcommittee investigation could identify no evidence that fusion centers contributed to uncovering a terrorist threat or disrupted an active terrorist plot.

And the American Civil Liberties Union argues that these centers are designed to allow the U.S. government to spy on Americans through data mining and "ambiguous lines of authority."

Now, you know I am strongly against any invasion of privacy, which is why I don't have a Facebook page and why I'm extremely careful of what I post anywhere online.

But is it really an invasion of privacy if fusion centers monitor social media and other public platforms? What I mean is if you post something on social media, you should realize that it could potentially be seen by millions of people, right?

And if someone is foolish enough to post highly personal — or illegal — information on social media, then really, it's their own fault if they get caught.

I saw a great quote the other day: "Dance like no one is watching; email like it may one day be read aloud in a deposition."

Basically, be careful what you announce to the world. Anything you say can and will be used against you.

A Spy's Eye View

Most people who work in the intelligence field will tell you that the biggest espionage threats in the U.S. are Chinese and Russian operatives. These two counties will do almost anything to learn our most guarded secrets — primarily our technological advances and military capabilities.

About 10 years ago, the FBI arrested a California man who had been stealing U.S. Navy secrets and providing them to the Chinese government. Chi Mak, who immigrated to the U.S. from China, was convicted of acting as an unregistered agent of a foreign government and exporting U.S. military secrets to China.

While he never admitted guilt (most spies won't), the FBI believes Mak was a trained Chinese intelligence officer who had been planted in the U.S. in the 1970s.

Keeping an Eye on Things

The investigation into Mak began in 2004 when the FBI received a tip that someone who worked for Power Paragon, a defense contractor, was sharing military secrets regarding power systems developed for the U.S. Navy.

The FBI began surveillance on Mak, who had worked for Power Paragon since 1988. For over a year, the FBI kept tabs on Mak, his wife and other family members. Agents also regularly went through the Mak's trash to find evidence to build their case. The FBI went so far as to conduct surveillance on the neighbors, even observing their bathroom tendencies during the night.

One evening when FBI agents knew Mak was out of town, a covert entry team went to Mak's home around midnight. They wanted to make sure none of the neighbors became suspicious, so they arrived driving the exact same vehicle Mak owned to blend in.

Once the entry team gained access to Mak's home, they photographed anything and everything in plain sight. They wanted all the evidence they could find, but obviously, they didn't want Mak to know they had been in his home. Based on key pieces of evidence discovered during the covert entry, the FBI was able to bring espionage charges against Mak.

The Eyes Have It

Clearly, the success of this case is owed to the intelligence gained from the FBI's painstaking surveillance efforts — planting wires, installing cameras, monitoring equipment... spending long days and even longer nights of listening, watching and waiting.

But what if there were a simpler solution? What if we had the technology that allowed an undercover agent to conduct surveillance using a contact lens in their eye?

That would give a whole new meaning to the phrase "keeping your eyes peeled."

It might sound far-fetched, but multiple technology companies including Google, Samsung and Sony have filed patents for contact lenses with various capabilities, including full imaging technology, storage capacity and the ability to connect to other wireless devices, like smartphones.

These lenses can transmit images, zoom, auto-focus and do most of the things a regular camera can do. To take a picture, all you have to do is blink. The lens can tell the difference between a natural blink and a longer blink, which will take a picture.

In the Eyes of the Law

Can you imagine how this technology would benefit our intelligence and law enforcement community? Think about the time and money it would save if agents had the ability to conduct surveillance or record information through their eyes?

We are continually seeing advancements in wearable technology that are changing the world we live in. A contact lens camera brings up privacy and security concerns. For example, it would be extremely difficult to prevent someone from recording private meetings. Nevertheless, this technology would be a valuable intelligence asset, enabling undercover agents to collect a large amount of information in the blink of an eye.

Printed in Great Britain
by Amazon